To Dianne, Lynn, and Mary—cherished critique partners and dear friends

—K. J.

For Fabian

—R. K.

BEACH LANE BOOKS
An imprint of Simon & Schuster Children's Publishing Division
1230 Avenue of the Americas, New York, New York 10020

Book design by Debra Sfetsios-Conover

The text for this book was set in Urbana.
The illustrations for this book were rendered digitally.
Manufactured in China
1125 SCP
First Edition
10 9 8 7 6 5 4 3 2 1
CIP data for this book is available from the Library of Congress.
ISBN 9781665975407
ISBN 9781665975414 (ebook)

# WHO NESTS HERE?

## Twenty-Four Extraordinary Animal Homes

*Written by* **Karen Jameson**

*Illustrated by* **Ramona Kaulitzki**

BEACH LANE BOOKS

New York Amsterdam/Antwerp London Toronto Sydney/Melbourne New Delhi

**WHO NESTS HERE?**

Whose baby bed?

Whose family home?

Whose winter shed?

Clues are waiting up ahead!

Who nests here?

How can you tell?

Twigs in branches,

Bumpy swell,

Leafy perch in which to dwell.

Who nests in trees?

Gray squirrels

Gall wasps
Apes

Who nests here?

How do you know?

Giant fortress,

Scrapes that show,

Turrets built when tide is low.

Who nests in mud?

African termites

Devil crayfish
Mudskippers

Who nests here?

What is your clue?

V-shaped marks,

Tracks in view,

Dragging tail-print—footprints, too.

Who nests in sand?

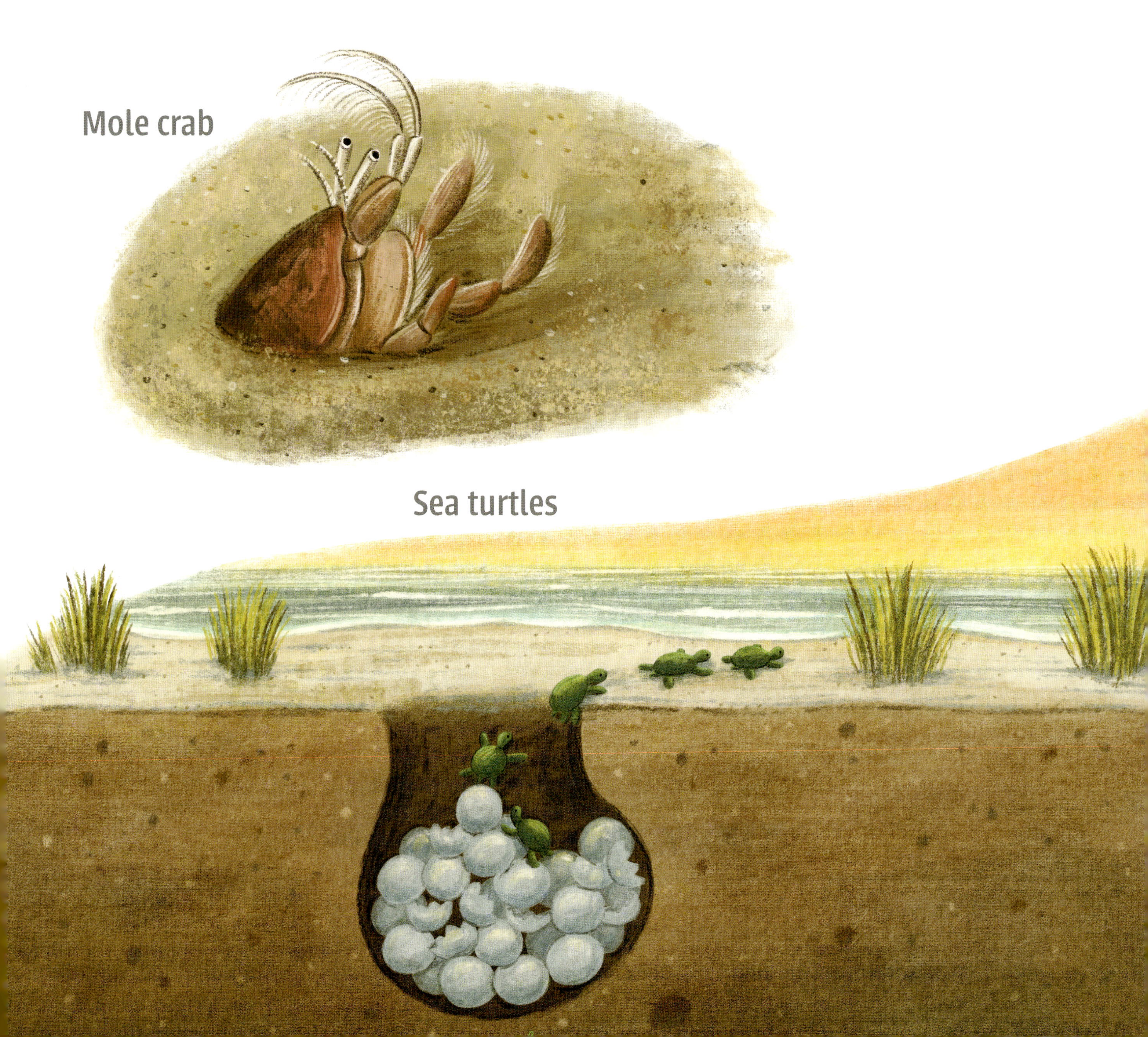
Mole crab
Sea turtles

Scorpion

Who nests here?

Can you see?

Lodge of sticks,

Gel debris,

Bubble nest where eggs may be.

Who nests in water?

Beavers

Stream-dwelling frog

Siamese fighting fish

Who nests here?

What hints are there?

Compost heap,

Leafy lair,

Pile of grasses stacked with care.

Who nests in plants?

American alligator

King cobra

Plains harvest mouse

Who nests here?

Look around!

Lots of small hills,

Littered ground,

Trees where gnaw marks can be found.
Who nests in dirt?

Naked mole rats
Foxes

Rabbits

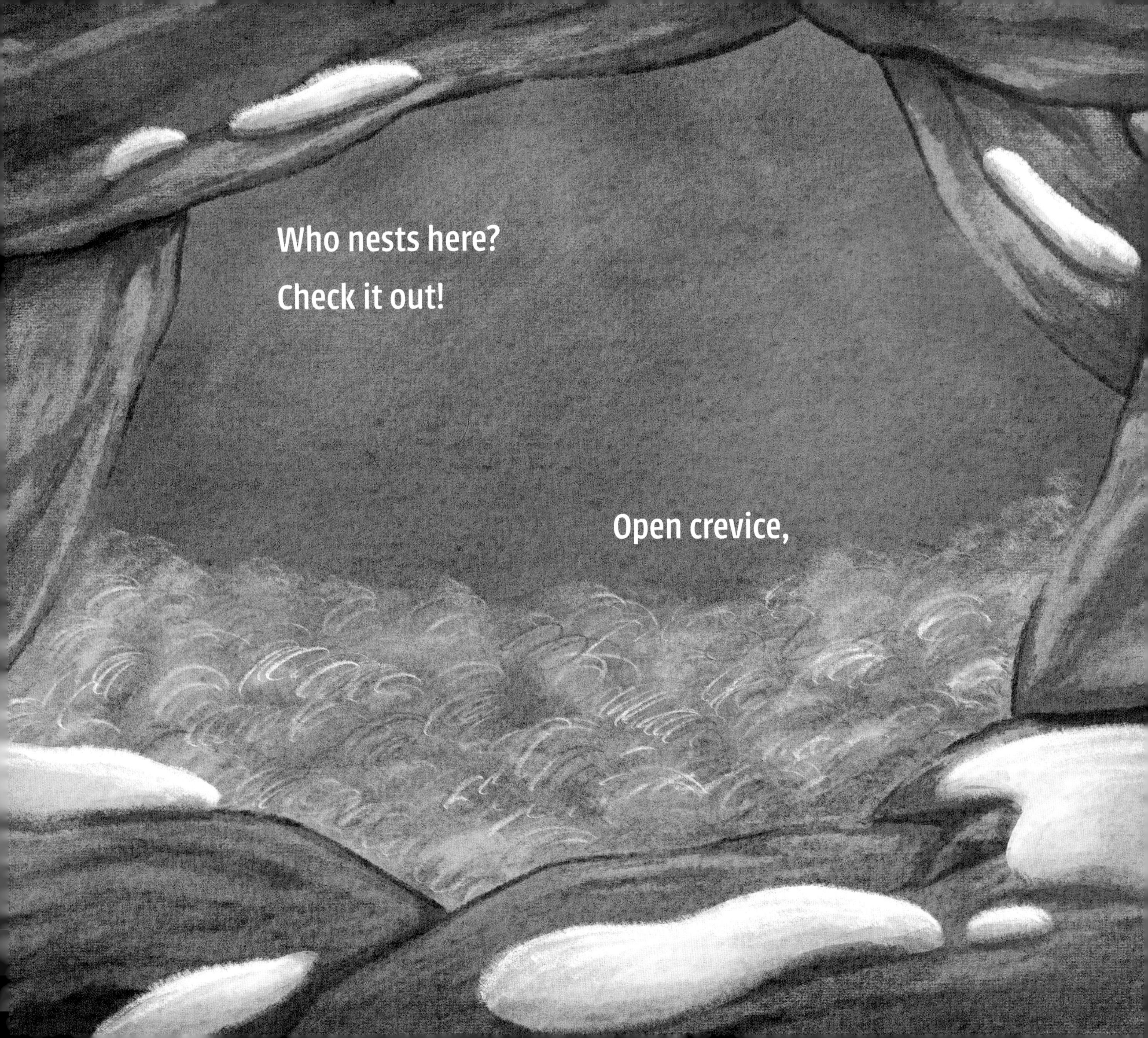

Who nests here?

Check it out!

Open crevice,

Bark scratched out,

Trail of crab shells strewn about.

Who nests in rocks?

Snow leopards

Black bear
Octopus

Who nests here?

Search left and right.

Tiny paw prints,

Mounds of white,

Foraged brush around the site.

Who nests in snow?

Vole
Polar bears

Woodchucks

# Who would have guessed? A world of nests!

Each special home a place to rest.

## Who Nests in Trees?

**Gray squirrels** build soccer-ball-sized twig nests called *dreys* and line them with moss and dry grass.

**Gall wasps** secrete a chemical that causes wartlike bumps called *galls* to grow on oak trees. Wasp grubs (babies) hatch inside them and feed on the centers.

**Apes**—such as orangutans, chimpanzees, gorillas, and bonobos—use leaves and twigs to build cradle-shaped nests in trees.

## Who Nests in Mud?

Some **African termites** spend years building huge mud fortresses, called *termitariums*, that tower up to sixteen feet tall.

If you see scrapes, grooves, and mud mounds along lake shores, they're probably the work of the **devil crayfish**.

Male **mudskippers** first dig burrows in mudflats, then leap around to attract females. If impressed, a female lays hundreds of eggs inside the burrow. The male spits out mud blobs to build small hills called *turrets* at the burrow's entrance. These turrets serve as a lookout post for guarding the eggs and overseeing the territory.

## Who Nests in Sand?

**Mole crabs**, a kind of sand crab, make V-shaped marks in the wet sand of the swash zone. How? With their antennae! As waves crash ashore, they burrow backward in a flash, using a second set of feather-like antennae to filter feed. Water passing through the antennae washes away the unwanted bits, allowing the crab to feast on its favorite prey. The Vs are quickly erased by the tide, as the busy crabs continuously move along the water's edge in search of their next meal.

A **sea turtle** mama kicks up a lot of sand while digging a pit for her eggs. After covering the eggs with sand, she returns to the ocean. Flipper tracks and crawl patterns are clues that the mama has left her nest. A few months later, hatchlings dig out of nests at night and scurry toward the sea.

**Scorpions** hide under desert sand to rest and stay cool during the daytime. A tail line between eight tiny footprints may signal a scorpion's burrow underneath the sand.

## Who Nests in Water?

**Beavers** build lodges made of piled twigs, branches, and mud in rivers and lakes. Tree stumps that have been ground to a point are a clear sign that a beaver has been hard at work.

**Stream-dwelling frogs** deposit masses of eggs in gelatinous material on underwater plants. This gel "nest" anchors the eggs so they won't wash downstream.

Male **Siamese fighting fish**, or betta fish, spend hours making mucus-coated air-bubble nests to attract females. Once a female lays eggs in the nest, the male guards them until they hatch.

## Who Nests in Plants?

The **American alligator** builds a large nest near water by piling up a combination of rotting and fresh vegetation, mud, and sticks. This compost heap bakes in the sun to become a warming incubator for the alligator's eggs.

The female **king cobra** makes a nest pile by pushing leaves and branches together with her head and body. Then she lies on top of the nest to guard her eggs.

**Plains harvest mice** use grasses to construct their nests, placing them in shrubs, in higher grasses, or on the ground.

## Who Nests in Dirt?

**Naked mole rats** dig underground tunnels full of chambers where they sleep, store food, and have their babies. When tunneling, naked mole rats kick out all the extra dirt through openings called *molehills*. This behavior is known as *volcanoing*.

**Fox** dens have many tunnels, entrances, and exits. Food remains, such as feathers and bones, are often littered around the area.

**Rabbit** teeth never stop growing. Gnawing on the roots above their underground burrow keeps the teeth filed down to a healthy length. Rabbits cover their burrow entrances with a patch of bare earth or grass to keep predators out.

## Who Nests in Rocks?

**Snow leopard** mamas give birth in well-hidden rocky dens. A bed lined with fur from a mama's underbelly keeps newborns warm.

**Black bears** make their dens in rock crevices, hollow trees or logs, or under tree roots. Shredded bark, paw prints, and leafy beds are telltale signs that a bear den is near.

The **octopus** holes up in rocky nooks on the ocean floor. A trail of crab shells may lead to this messy eater's den.

## Who Nests in Snow?

**Voles** tunnel and make nests in the subnivean zone—the blanket of snow nearest the ground. Look for little round tunnel openings and tiny, faint paw prints.

Pregnant **polar bears** carefully choose a site with deep snow and dig their birthing den there. The dens are hard to spot, but you may see a snow mound with a single covered entrance.

**Woodchucks**, also called groundhogs, nest in underground burrows during the long, snowy winter. Before hibernating, they fatten up by gorging on plenty of local brush and plants.

## More about Nesting

From the tiniest insect to the largest bear, animals of all sizes and kinds make nests. For some it's a place to sleep, for others a spot to raise a family or hide from predators. For social creatures, like termites or naked mole rats, a nest is a little city unto itself, where the community lives and works together. Still others, like squirrels and groundhogs, hibernate all winter in the same place while apes build a new nest in a different place every night. Protecting animal homes, such as nests, is key to conserving our diverse wildlife in every habitat on earth. By doing so, we save the amazing species that live in our natural world.

## Nest Safety

Always respect animals and their homes. Observe them from a distance. Don't touch or disturb them. Don't poke sticks in a nest or put your hands inside one. Never remove eggs from a nest or move a nest to another site. Animals will do what they must to protect their homes and families, so don't give them any reason to be fearful. We can all stay safe and enjoy nature when we respect one another.

## Nesting Site Conservation

To survive, animals need a healthy habitat to provide food, water, shelter, and other natural resources. But sadly, some animals are losing their homes to climate change, deforestation, drought, overhunting, pollution, new construction, and many other threats. A number of animals have been so greatly affected that they're on the endangered species list. Nesting sites are especially important to conserve, or protect, as they are the nurseries of the wild. New generations of babies help ensure the future of each species.

## What Can You Do to Help?

You can keep nesting sites safe by supporting national parks and other protected wildlife spaces. Ask an adult to alert authorities, such as rangers and animal control, to new nests that you find, so they can rope them off and put up signs. When camping or walking on trails, pick up your trash to keep the paths and waterways clean. Leave rocks, shells, pine cones, and other natural elements as they are. They may be someone's home. Learn about conservation and share what you know with your friends and family.